IT'S A
BEAUTIFUL
WORLD

––––––––

Have you been to Yosemite Valley? That's a place that can make you catch your breath: the scale of the rock faces, the delirious perfection of the valley floor's meadows. It makes your head swim, a recognition of beauty so powerful, you just … stop.

The world is full of places like that. But we don't see them every day and sometimes we need to be reminded that they are there. This book gathers together photography of some of the world's most extraordinary places to share the wonder they bring.

Being in a beautiful place is more than observing what's before your eyes. The experience of beauty is an emotional one, brought about through context; there's the people you're with, or the people you're not with; there's your mood, where you are in your life. All these factors come together to make the response to a place unique and personal, and different every time you visit.

When you look at a photograph of a beautiful place, that context disappears. We thought about that fact when we were making this book. We wanted to create a context that helped bring some of those connections to bear. The images here are arranged in chapters that reflect an aspect of life. As you turn the pages we hope that pondering the relationship between each image and the life stage with which it's been connected will add another dimension to your experience.

The images in this book will take you to places far and wide, the kinds of places that you might never visit but that you can perhaps put on that "If" list we all have tucked away. These places are surprising, remarkable, remote, familiar … dive in and marvel over the undeniable fact; it *is* a beautiful world.

~ ORIGINS ~

Sequoia National Park › California, USA

Origins

Stone pinnacles at Cavusin › Cappadocia, Turkey

Moai (statues) › Easter Island, Pacific Ocean

Origins

Dawn at Canyonlands National Park › Utah, USA

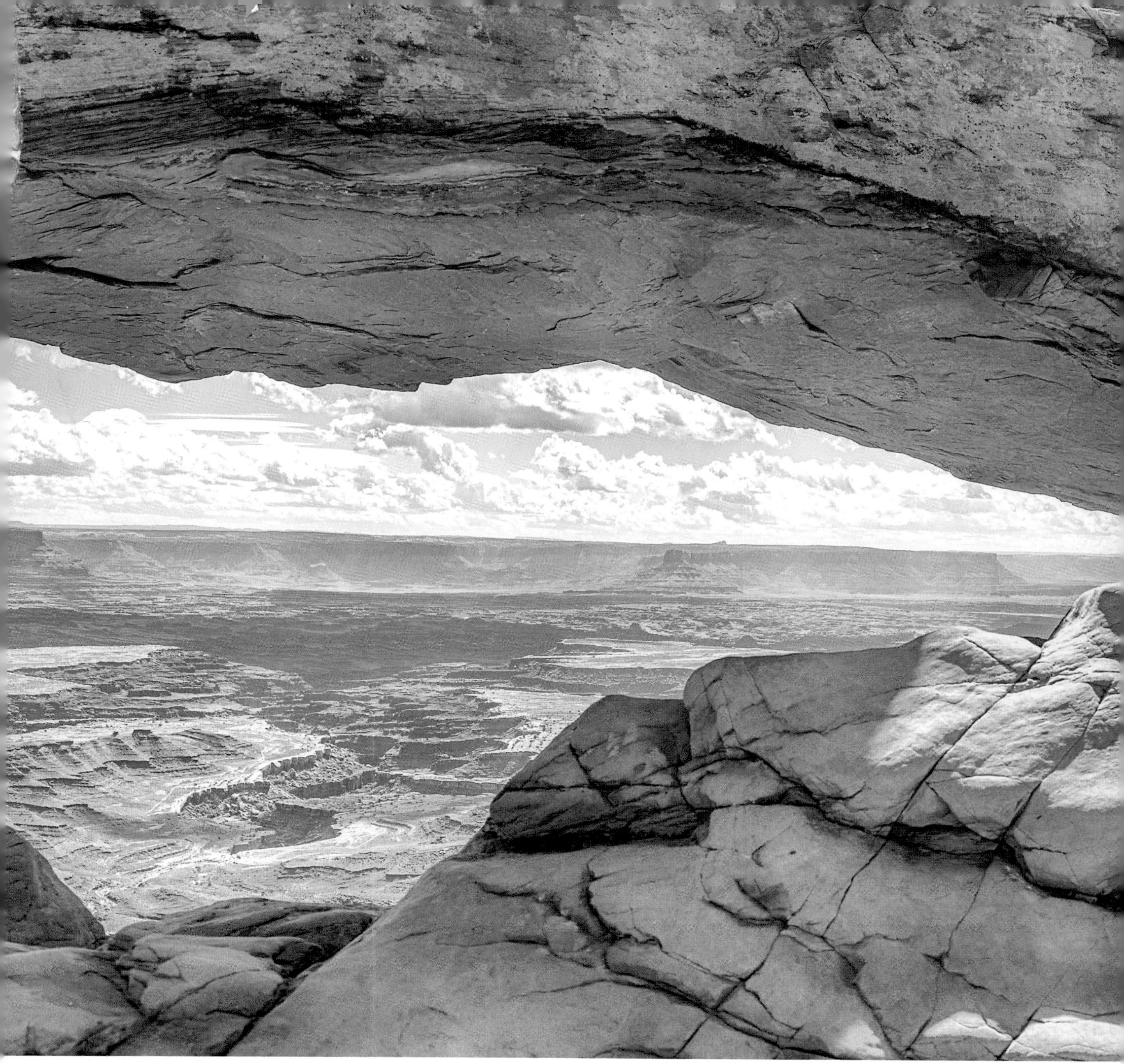

Ait Ben Haddou › Morocco

Rice terraces at Yuangyuan › Yunnan, China

The Rwenzoris › Uganda

Origins

Plitvice Lakes National Park › Croatia

Avenue of the Baobabs › Madagascar

Origins

Amboseli National Park sunrise › Kenya

Fly geyser, Black Rock Desert › Nevada, USA

The Bungle Bungles › Purnululu National Park, Western Australia

Halema'u ma'u crater › Hawaii, USA

Lava at Kalapana › Hawaii, USA

Origins

Kirkjufellsfoss › Iceland

∼ NOURISH ∼

Grand Canyon National Park › Arizona, USA

Nourish

South Downs National Park › West Sussex, England

Okavango River › Botswana

The Sardine Run › Eastern Cape, South Africa

Saint-Émilion › France

Nourish

Elephant › Masai Mara, Kenya

Wildebeest › Serengeti National Park, Tanzania

Steller's sea eagles › Kamchatka, Russia

Rice terraces › Longsheng, China

Nourish

Umm al-Maa lake › Ubari Sand Sea, Libya

Nourish

Incahuasi Island › Salar de Uyuni, Bolivia

Yulong River › Guangxi Zhuang, China

Buffalo at Yellowstone National Park › Wyoming, USA

Lavender on the Plateau de Valensole › Alpes de Haute-Provence, France

Lyth Valley in the Lake District › Cumbria, England

Humpback whales in Chatham Strait › Alaska, USA

Nourish

Puffins › Fair Isle, Scotland

Kalsoy Island › The Faroe Islands

~ UNTAMED ~

Porthcawl › Wales

Untamed

The Na Pali coast of Kaua'i › Hawaii, USA

Surfing Pipeline at O'ahu › Hawaii, USA

Bárðarbunga volcano › Iceland

Chamarel waterfall › Mauritius

Freshwater lagoons › Lençóis Maranhenses National Park, Brazil

Iceberg arch › Antarctica

Ta Prohm temple › Angkor Wat, Cambodia

Untamed

Lake Baikal › Siberia, Russia

Svartifoss › Iceland

The Rub al Khali › Oman

Icelandic horses › Iceland

Condors › Colca Canyon, Peru

Untamed

A dust devil › Serengeti National Park, Tanzania

A lionness › Serengeti National Park, Tanzania

Untamed

Hoh rainforest › Olympic National Park, Washington, USA

Torres del Paine National Park › Patagonia, Chile

Untamed

Pico Cão Grande › São Tomé and Príncipe

Cape Tribulation › Queensland, Australia

Cape Raoul › Tasmania, Australia

Great White shark › Gaudalupe Island, Mexico

Avalanche › Rhône-Alpes, France

~ COMMUNITY ~

Ogimachi village › Gifu, Japan

Community

Chefchaouen › Morocco

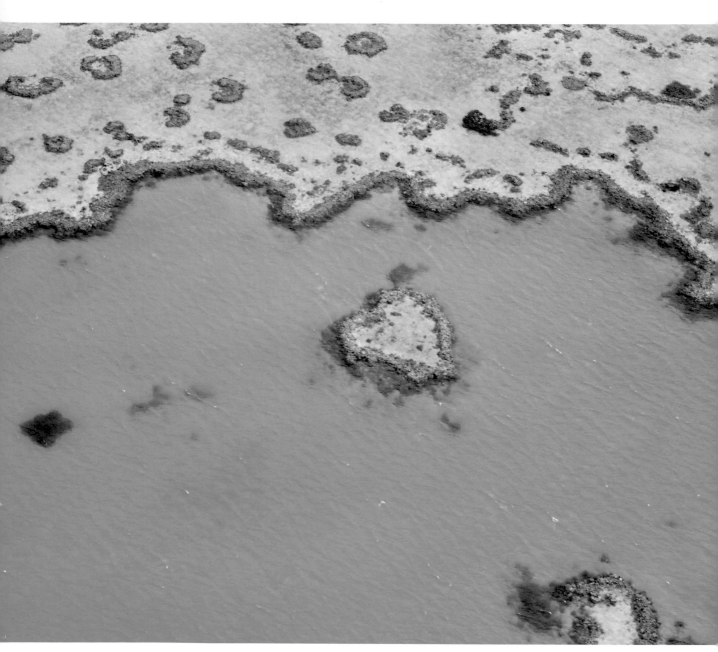

The Great Barrier Reef › Queensland, Australia

Community

Manhattan › New York, USA

Water buffalo › Ban Gioc, Vietnam

Red deer in Richmond Park › London, England

King penguins › Antarctica

Green turtles › Galápagos Islands, Ecuador

Halong Bay › Gulf of Tonkin, Vietnam

Monarch Butterfly Biosphere Reserve › Michoacán, Mexico

Manarola town in Cinque Terre › Liguria, Italy

A water hole in Etosha National Park › Namibia

Djenné › Mali

Sun City › Arizona, USA

Vidigal favela › Rio de Janeiro, Brazil

Reine village › Lofoten Islands, Norway

~ CELEBRATION ~

Ceann Hulavig › Isle of Lewis, Scotland

Mani Rimdu festival › Sagarmatha, Nepal

Cherry blossom › Yuantouzhu, China

San Andrés Apóstol cemetery › Mixquic, Mexico

Wandering albatrosses › South Georgia Island, Atlantic Ocean

Holi festival › India

The Sardine Run › Eastern Cape, South Africa

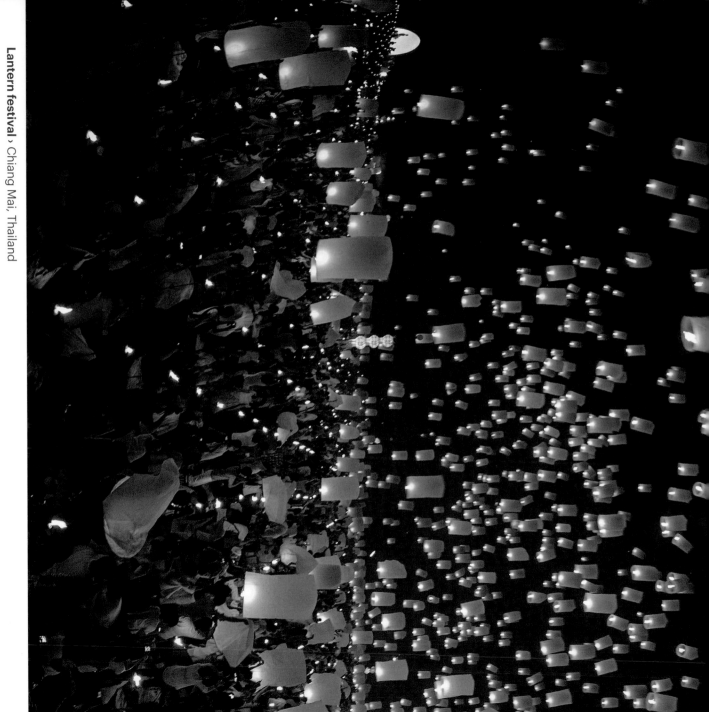

Lantern festival › Chiang Mai, Thailand

Snow geese › Canada

Polar bear › Svalbard, Norway

Celebration

Vineyards in Greve › Tuscany, Italy

Celebration

Danum Valley Conservation Area › Borneo, Malaysia

Celebration

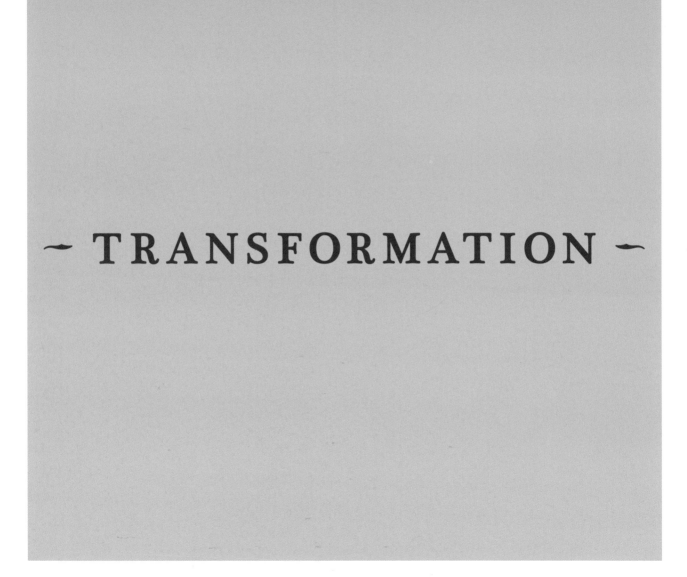

~ TRANSFORMATION ~

The Austfonna ice cap › Svalbard, Norway

Transformation

The Dolomites › South Tyrol, Italy

Transformation

Monument Valley Tribal Park › Arizona–Utah, USA

Reed Flute Cave › Guangxi, China

Transformation

Yellowstone National Park › Wyoming, USA

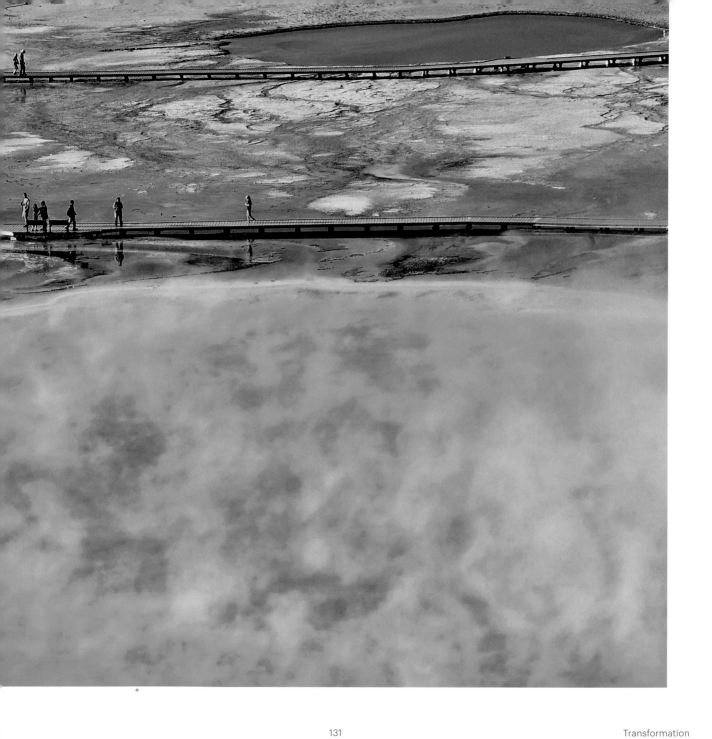

Transformation

Wildflowers in the Atacama Desert › Chile

Transformation

The Black Forest › Baden-Württemberg, Germany

Drakes Passage › The Southern Ocean, Antarctica

Transformation

The aurora borealis › Reine, Norway

Transformation

Zhangye Danxia Landform Geological Park › Gansu, China

Fall leaves › Vermont, USA

Bora Bora › French Polynesia, Pacific Ocean

A supercell storm near Severy › Kansas, USA

Horseshoe Bend › Arizona, USA

Transformation

Iceberg in Grandidier Channel › Pleneau Island, Antarctica

Slot canyon › Utah, USA

Transformation

Eyjafjallajökull volcano › Iceland

Grand Canyon › Arizona, USA

Transformation

Skógafoss › Iceland

Transformation

~ SPACE ~

Lighthouse Reef › Belize

Milford Sound › Fiordland National Park, New Zealand

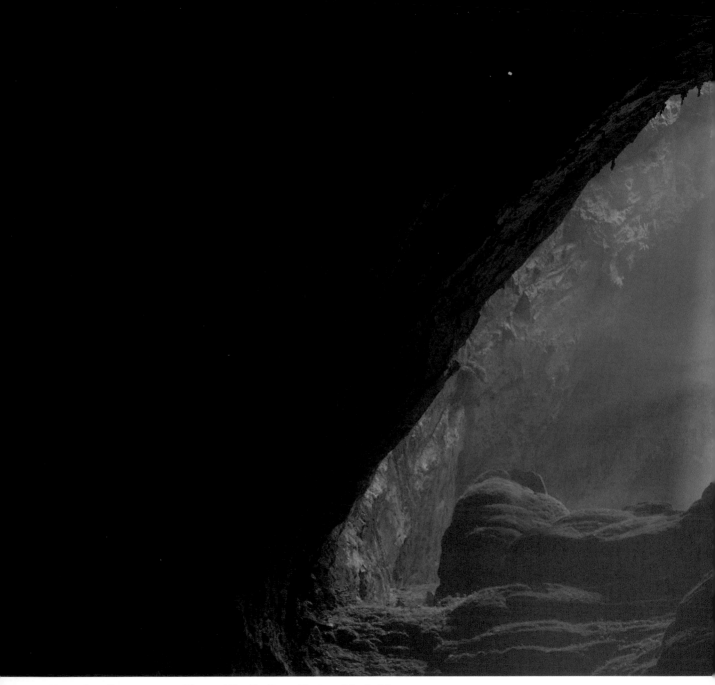

Hang Son Doong cave › Phong Nha-Ke Bang National Park, Vietnam

Seljalandsfoss › Iceland

A cenote near Valladolid › Yucatán, Mexico

Space

Meteorite crater at Gosse Bluff › Northern Territory, Australia

Picos de Europa › Asturias, Spain

Hammerhead sharks › Galápagos Islands, Ecuador

Edgerton Highway › Alaska, USA

Oneonta Creek, Columbia River Gorge › Oregon, USA

Victoria Harbour › Hong Kong

Space

McDonald Observatory › Texas, USA

Sunset › Antarctic Peninsula

Space

White Sands National Park › New Mexico, USA

Sossusvlei sand dunes › Namibia

Space

Hadrian's Wall › Northumberland, England

~ HARMONY ~

Harmony

Cypresses in the Val d'Orcia › Tuscany, Italy

Pura Ulun Danu Bratan temple › Bali, Indonesia

The Devil's Marbles in Tennant Creek › Northern Territory, Australia

Harmony

Red-crowned cranes › Hokkaido, Japan

Harmony

Triglav National Park › Slovenia

Harmony

Reindeer › Sweden

Mustering horses › Xilin Gol, Mongolia

Harmony

Moraine Lake in Banff National Park › Alberta, Canada

Highland cattle › Scotland

The Cuillins › Isle of Skye, Scotland

Harmony

The Kimberley coast › Western Australia

A quiver tree › Kalahari, Namibia–South Africa

The Kamakura Bamboo Garden › Tokyo, Japan

Harmony

Lamayuru Monastery › Ladakh, India

Rice paddies › Bali, Indonesia

Tianmen mountain road › Hunan, China

Harmony

Wildebeest migration › Masai Mara, Kenya

Monumental

Victoria Falls › Zambia–Zimbabwe

Bryce Canyon National Park › Utah, USA

The Gasherbrum massif › Karakorum, Pakistan

Monumental

Mount Bromo › East Java, Indonesia

Angkor Wat › Siem Reap, Cambodia

Baobab tree › Sri Lanka

Monumental

Mt Fitzroy in Los Glaciares National Park › Patagonia, Argentina

Monumental

Whale shark › Western Australia

Ningaloo National Park › Western Australia

Monumental

The Palace of Westminster › London, England

Monument Valley › Arizona–Utah, USA

Monumental

Elephants › Okavango Delta, Botswana

Monumental

Millau viaduct › Midi-Pyrénées, France

Yosemite National Park › California, USA

Strokkur geyser › Iceland

Limestone pinnacles at Wulingyuan › Hunan, China

Monumental

~ ETERNAL ~

Angel Falls › Venezuela

Eternal

The Royal Tomb › Petra, Jordan

Teahupo'o › Tahiti

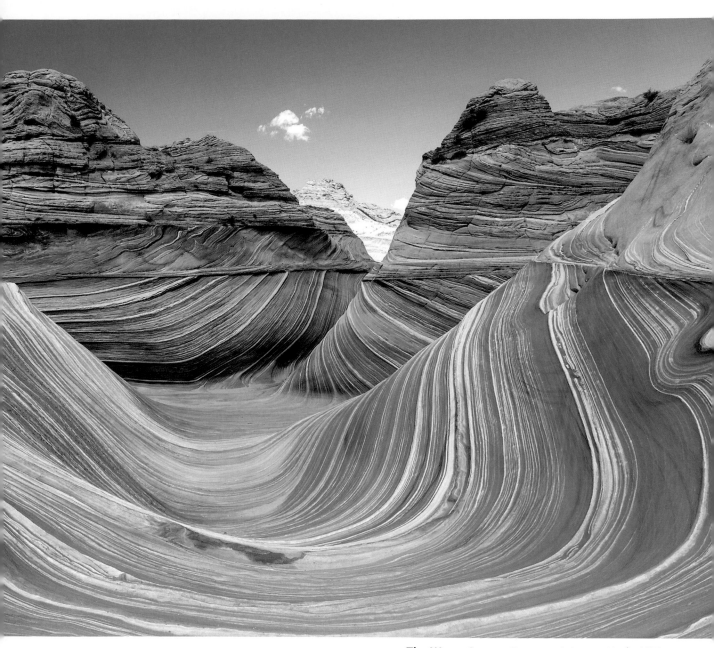

The Wave, Coyote Buttes › Arizona–Utah, USA

Eternal

Uluru › Northern Territory, Australia

Taj Mahal › Uttar Pradesh, India

Eternal

Zabriskie Point, Death Valley › California, USA

Eternal

Terracotta army, Xi'an › Shaanxi, China

Shiprock › New Mexico, USA

Eternal

The Temple of Saturn › Rome, Italy

Bristlecone pine tree › California, USA

Eternal

Bondi › New South Wales, Australia

Bagan temples › Mandalay, Myanmar

West Elk Wilderness › Colorado, USA

Eternal

~ INDEX ~

Sequoia National Park › California, USA The General Sherman Tree, a sequoia found in California's Sequoia National Park, is the world's largest tree. It is more than 83m tall and has a diameter of 7.7m. It's thought to be around 2000 years old.

7

David Clapp | Getty Images

The Milky Way › The Dolomites, Italy Home to our solar system, the mighty Milky Way contains more than 200 billion stars and perhaps almost as many planets – not least of which is Earth.

8

Anita Stizzoli | Getty Images

Sulphuric lakes at Dallol › Danakil Desert, Ethiopia The ancient volcanic crater in Dallol creates amazing heat (it is said to be the hottest place on earth), colourful, acidic ponds, mountains of sulphur, and an out-of-this-world landscape.

9

Lukas Bischoff Photograph | Shutterstock

Stone pinnacles at Cavusin › Cappadocia, Turkey Cappadocia is a geological oddity of honeycombed hills and towering boulders that were fashioned through volcanic ash and moulded by millennia of rain and river flow.

10

IgorZh | Shutterstock

Moai › Easter Island, Pacific Ocean The monolithic human figures on Easter Island were carved by the Rapa Nui people sometime between the years 1250 and 1500. The tallest statue is almost 10m high.

11

Volanthevist | Getty Images

Canyonlands National Park › Utah, USA Dawn's warm light hits the rock formations of Canyonlands. The arches, bridges, needles, spires and craters of Utah's largest park were carved by the Colorado River.

12

canadastock | Shutterstock

Ait Ben Haddou › Morocco This fortified village, in a water-rich valley, lies on a former trans-Sahara caravan route to Marrakech. It has been a UNESCO World Heritage Site since 1987.

14

Tom Camp | Shutterstock

Rice terraces at Yuanyuang › Yunnan, China In the southwest of China lies this fertile plateau of rice paddies. Agriculture in east Asia began 10,000 years ago, and prompted people to settle in one area.

15

JohnPhotoSiam | Shutterstock

Grand Canyon in the Blue Mountains › New South Wales, Australia The Blue Mountains National Park, 80km west of Sydney, comprises waterfalls, canyons and rainforest. An ancient tree species, the Wollemi pine, was found in a gorge here.

16

Ted Mead | Getty Images

The Rwenzoris › Uganda Rwenzori means 'rainmaker' in the local language and this range, on the border of Uganda and the Democratic Republic of the Congo was thought to be the source of the Nile. They're also known as the Mountains of the Moon.

17

Nikolai Link | Shutterstock

Plitvice Lakes National Park › Croatia This Unesco World Heritage–listed park is made up of interlinked and cascading lakes, caves and forest. The colours of the lakes range from azure to green, blue and even grey.

18

Kelly Cheng | Getty Images

Avenue of the Baobabs › Madagascar These bulbous trees, known in Malagasy as 'mother of the forest', are up to 2800 years old and were once part of thick forest. These tall baobabs are all that remain.

19

Justin Foulkes | Lonely Planet

Amboseli National Park sunrise › Kenya A new day in Amboseli National Park and more challenges await for the inhabitants of one of Africa's best places to see wildlife. But in 2020 more than 170 elephant calves were born.

20

Jane Rix | Shutterstock

Fly geyser, Black Rock Desert › Nevada, USA While drilling for water in 1916, farmers struck a geothermal pocket of water; dissolved minerals accumulated to create colourful growths around the geyser.

21

Lukas Bischoff Photograph | Shutterstock

The Bungle Bungles › Purnululu National Park, Western Australia The beehive-like landscape of the Bungle Bungles was formed by millions of years' worth of wind and rain. The distinctive mounds are striped with orange, black and grey bands.

22

Eric Middelkoop | Shutterstock

Halema'u ma'u crater › Hawaii, USA Halema'u ma'u crater is part of Kilauea volcano, which has erupted repeatedly since 1823, the year of the volcano's first well-documented eruption. Its lava splashes into the ocean and creates a new coastline.

24
Matt Munro | Lonely Planet

Lava at Kalapana › Hawaii, USA In the late 1980s and early 1990s lava flow invaded and destroyed the Hawaiian town of Kalapana. The landscape continues to change from day to day.

25
Henner Damke | Shutterstock

Kirkjufellsfoss › Iceland Kirkjufellsfoss – and the mountain Kirkjufell behind it – are popular stops on Iceland's Snæfellsnes peninsula. The shape of the mountain is said to have inspired the architect of Reykjavík's Hallgrímskirkja.

26
Jonathan Gregson | Lonely Planet

Grand Canyon National Park › Arizona, USA The Grand Canyon is vast and nearly incomprehensible in age – it took six million years for the canyon to form and some rocks exposed along its walls are two billion years old.

29
Matt Munro | Lonely Planet

South Downs National Park › West Sussex, England An area of outstanding beauty, the South Downs is England's newest national park. The area is home to more than 110,000 people and is composed of working farmland and chalk hills.

30
Justin Foulkes | Lonely Planet

Okavango River › Botswana The Okavango changes with the seasons as flood waters ebb and flow, creating islands, river channels and pathways for animals that move this way and that at the waters' behest.

31
Cedric Favero | Getty Images

The Sardine Run › Eastern Cape, South Africa During the annual migration of sardines, millions of the fish head north along the coast of South Africa, creating a feeding frenzy along the way. The migration occurs from May to July.

32
Dmitry Miroshnikov | Getty Images

Saint-Émilion › Gironde, France One of the most famous names in French wine-making, grapes have been grown around this village near Bordeaux for 2000 years. Today it is also an UNESCO World Heritage Site.

34
Justin Foulkes | Lonely Planet

Elephant › Masai Mara, Kenya The sweeping savannah of the Masai Mara is the place for the world's most spectacular display of wildlife. The drama is at its most intense in August, the start of the wildebeest migration.

36
nwdph | Shutterstock

Wildebeest › Serengeti National Park, Tanzania Following the rains of East Africa, more than a million wildebeest cross the Serengeti's plains to graze and raise their young.

37
Jonathan Gregson | Lonely Planet

Steller's sea eagles › Kamchatka, Russia Steller's sea eagles thrive on a diet of salmon, and in the shallow, icy streams of Kamchatka they often find the fish frozen near the surface.

Page 38
Ondrej Prosicky | Shutterstock

Rice terraces › Longsheng, China Terraced paddy fields wind up from the riverside to the mountain top in a feat of farm engineering that allows the communities of Longsheng to harvest rice in a mountainous area.

40
KingWu | Getty Images

Umm al-Maa lake › Ubari Sand Sea, Libya Surrounded by sand, the Umm al-Maa (or Mother of Water) is a refreshing oasis in the Libyan desert. The lake is fed by natural springs, which have slowly been drying up.

42
Patrick Poendl | Shutterstock

Incahuasi Island › Salar de Uyuni, Bolivia The hilly outpost of Incahuasi Island is covered in Trichocereus cacti and surrounded by a flat white sea of hexagonal salt tiles. The salty expanse is an evocative and eerie sight.

44
Delpixel | Shutterstock

Yulong River › Guangxi Zhuang, China The Yulong and Li River area is renowned for classic images of mossy-green jagged limestone peaks, bubbling streams, wallowing water buffalo and cormorant fishing.

45
Mark Read | Lonely Planet

Buffalo at Yellowstone National Park › Wyoming, USA More than 3000 buffalo roam throughout the Yellowstone National Park in Wyoming; they are among the last wild buffalo in the United States.

46

Don Mammoser | Shutterstock

Plateau de Valensole › Alpes de Haute-Provence, France Lavender from the fields of the Plateau de Valensole is made into lavender oil, honey, soap and scented sachets. The lavender fields usually bloom in July.

48

emperorcosar | Shutterstock

Lyth Valley in the Lake District › Cumbria, England The unspoilt Lyth Valley is tucked in a hidden corner of Cumbria, where trees are laden with fruit and rolling hills are the most magnificent green.

49

221A | Getty Images

Humpback whales in Chatham Strait › Alaska, USA The humpback whales in Chatham Strait spend their summers feeding on the healthy supply of krill and small bait fish that live in the frigid waters.

50

Gudkov Andrey | Shutterstock

Puffins › Fair Isle, Scotland Each year during spring and summer, puffins set up home on Fair Isle, between the islands of Shetland and Orkney. In 2019, Fair Isle's bird observatory burned down and is being rebuilt.

52

Finn Beales | Lonely Planet

Kalsoy Island › The Faroe Islands In the north of this archipelago, Kalsoy's promontories bear the brunt of the Atlantic weather. The Kallur lighthouse warns shipping away. The island's name means 'man'; 'woman island', Kunoy, is nearby.

53

Justin Foulkes | Lonely Planet

Porthcawl › Wales For much of the year, Porthcawl is a sedate seaside resort on the south coast of Wales with a sandy beach and a promenade. But when autumn storms arrive, a personality change takes place.

55

Steved_np3 | Shutterstock

The Na Pali coast of Kaua'i › Hawaii, USA For six million years the Pacific Ocean's waves have broken on Kaua'i's volcanic shores. The Na Pali coast in the northwest is a State Wilderness Park, attracting hikers and kayakers.

56

Matt Munro | Lonely Planet

Surfing Pipeline at O'ahu › Hawaii, USA Surfing was redefined in Hawaii in 2000 when riders began to be towed into monstrous waves by jet skis. Pipeline is a reef break on O'ahu's north shore.

57

Phillip B. Espinasse | Shutterstock

Bárðarbunga volcano › Iceland This stratovolcano, smouldering beneath a glacier is Iceland's second highest mountain and a reminder of the geological forces at play in our world. It last erupted in 2015.

58

Nathan Mortimer | Shutterstock

Chamarel waterfall › Mauritius The St Denis river cascades for 83m (272ft) over the edge of the Chamarel plateau in the southwest of the island. You can swim, surrounded by jungle, at its base.

59

Jonathan Stokes | Lonely Planet

Freshwater lagoons › Lençóis Maranhenses National Park, Brazil Freshwater lagoons are formed when these low-lying sand dunes are flooded during the Amazon basin's rainy season. But despite the water, very little vegetation survives here.

60

Erica Catarina Pontes | Shutterstock

Iceberg arch › Antarctica One of the reasons why icebergs float is that they contain a lot of air. Another reason is that they're formed from freshwater, which is less dense that salty water.

62

kajophotography.com | Getty Images

Ta Prohm temple › Angkor Wat, Cambodia Ta Prohm was built 800 years ago but was abandoned in the 17th century after the fall of the Khmer empire. Over subsequent centuries the jungle reclaimed the ruins.

63

Paul Biris | Getty Images

Lake Baikal › Siberia, Russia The world's deepest freshwater lake is perhaps also its oldest, at up to 30 million years of age. It's fed by 330 rivers, which descend off the surrounding mountains.

64

Tandemich | Shutterstock

Svartifoss › Vatnajökull National Park, Iceland The basalt columns of this waterfall in southeast Iceland inspired Reykjavík's Hallgrímskirkja, Iceland's largest church.

65
Creative Travel Projects | Shutterstock

The Rub Al Khali › Oman Better known, perhaps, as the Empty Quarter, this expanse of sand dunes, silence (and, beneath the surface, vast quantities of oil) comprises part of the Arabian desert and is largely unexplored.

66
Justin Foulkes | Lonely Planet

Icelandic horses › Iceland Occupying the space between wild and domesticated, Iceland's iconic purebred horses roam free every summer and are rounded up at the end of the season.

68
Jonathan Gregson | Lonely Planet

Condors › Colca Canyon, Peru In one of the world's deepest canyons, Andean condors use their 3m-wide wingspan to soar on rising currents of air, watching out for carrion below.

69
Galyna Andrushko | Shutterstock

A dust devil › Serengeti National Park, Tanzania The national park covers 14,750 sq km (5700 sq miles) of savannah, forest and grasslands. Human habitation is forbidden, except for park authority employees.

70
Jonathan Gregson | Lonely Planet

A lionness › Serengeti National Park, Tanzania Lions were first protected in this region in the 1920s after hunting had made them scarce. They share the land with cheetahs, hyenas, leopards and East African wild dogs.

71
Jonathan Gregson | Lonely Planet

Hoh rainforest › Olympic National Park, Washington, USA Rain falls frequently in the Hoh rainforest on the Olympic Peninsula, creating a thick forest of coniferous and deciduous trees draped in mosses and ferns.

72
Matt Munro | Lonely Planet

Torres del Paine National Park › Patagonia, Chile The granite pillars of Torres del Paine (Towers of Paine) dominate South America's most famous national park. In 2014, melting glaciers revealed fossils of Ichthyosaurs that swam here 200 million years ago.

73
Philip Lee Harvey | Lonely Planet

Pico Cão Grande › São Tomé and Príncipe This volcanic plug can be found in the south of São Tomé island, surrounded by snake-infested jungle. But its 1266ft spire (386m) still attracts a few brave climbers each year.

74
Justin Foulkes | Lonely Planet

Cape Tribulation › Queensland, Australia Cape Tribulation, in the Daintree National Park of tropical Queensland in the far northeast of Australia, is where surfaced roads cease and a 4WD is required to go any further.

76
Ewen Bell | Lonely Planet

Cape Raoul › Tasmania, Australia On the south coast of Tasmania, Australia's southernmost island state, these dolerite columns look out towards Antarctica. The cape makes a good day hike, alongside the Three Capes Track.

77
Catherine Sutherland | Lonely Planet

Great White shark › Gaudalupe Island, Mexico From the day they hatch, Great White sharks have to remain in perpetual motion so oxygenated water flows over their gills. An adult will eat 11 tons of food per year.

78
Ken Jones | Shutterstock

Avalanche › Rhône-Alpes, France Avalanche prediction is more art than science, with snow types and weather patterns causing unstable layers of snow. When the snow does let go, it can fall at 130km/h.

79
Marco Maccarini | Getty Images

Ogimachi village › Gifu, Japan In Shirakawa-go, deep in central Japan's mountains, the traditional farmhouses had steep roofs to withstand the heavy snows. The loft was used to cultivate silk worms.

81
Agustin Rafael C Reyes | Getty Images

Chefchaouen › Morocco Perched beneath the raw peaks of the Rif, Chefchaouen is one of the prettiest towns in Morocco, an artsy, blue-washed mountain village that feels like its own world.

82
Ken YEW (Singapore) | Getty Images

The Great Barrier Reef › Queensland, Australia The world's most extensive coral reef system supports 1500 species of fish and 4000 types of mollusc, each dependent on another. It's the most biodiverse of Unesco's World Heritage sites.

83
Matt Munro | Lonely Planet

Manhattan › New York, USA At almost 40,000 people per sq km in 1910, Manhattan's population density was higher then than in 2010. But at today's median price of US$800,000 for an apartment, the cost of property has gone up.

84
Nikada | Getty Images

Water buffalo › Ban Gioc, Vietnam The waterfall at Ban Gioc, on Vietnam's border with China, has long been the focal point of the local community, somewhere livestock can be washed and watered.

86
HNH Images | Getty Images

Red deer in Richmond Park › London, England Introduced in the 16th century to entertain royal hunting parties, red deer still roam relatively freely in Richmond Park. The rutting season, when males compete, starts in September.

88
arturasker | Shutterstock

King penguins › Antarctica King penguins congregrate to raise their chicks, herding the young birds together to protect them from the cold. The chicks take up to 10 months to fledge.

90
DLILLC | Corbis

Green turtles › Galápagos Islands, Ecuador To lay their eggs, Green turtles often return to the exact beach where they hatched. Just 1% of hatchlings will reach maturity to repeat the cycle.

91
Longjourneys | Shutterstock

Halong Bay › Gulf of Tonkin, Vietnam Evidence suggests Halong Bay's 500-million-year-old limestone islands have been a home to people for 20,000 years; today, four fishing villages support up to 2000 people.

92
Matt Munro | Lonely Planet

Monarch Butterfly Biosphere Reserve › Michoacán, Mexico Each winter, a billion butterflies flap across North America to this reserve in Mexico. The journey exceeds the insect's lifespan; no one knows how different generations return to the same location.

94
Roberto Michel | Shutterstock

Manarola town in Cinque Terre › Liguria, Italy The five villages of the Italian Riviera date back centuries, when locals grew grapes, caught fish and dreaded pirate attacks. Today sightseers are the source of income.

95
Gianluca Giardi | Getty Images

A water hole in Etosha National Park › Namibia Where there's water, animals in Africa can't be fussy about the company they keep; giraffes, zebras, kudus and namaqua sandgrouse take a drink in arid Etosha National Park.

96
Westend61 | Getty Images

Djenné › Mali In the city of Djenné, Mali, the Great Mosque is one of the world's greatest examples of adobe architecture. The entire community maintains the structure at an annual festival.

98
Ricardo Canino | Shutterstock

Sun City › Arizona, USA Launched in the 1960s, Sun City is a modern town popular with retirees. Creating a sense of community in ever-increasing suburbia is a challenge for urban planners.

99
gokturk_06 | Shutterstock

Starlings › The Netherlands A murmuration is a giant flock of starlings that ebbs and flows and swoops as one, above the birds' communal roosting site. Each bird tracks the movement of its seven closest neighbours, enabling the flock to fly as one.

100
Albert Beukhof | Shutterstock

Vidigal favela › Rio de Janeiro, Brazil This neighbourhood overlooks Ipanema beach. Once plagued by gang wars, it is gradually gentrifying, with an influx of incomers bringing new pressures.

102
luoman | Getty Images

The village of Reine › Lofoten Islands, Norway The fishermen's cottages on the Lofoten Islands, an Arctic archipelago, provided shelter between winter fishing trips; the catch was dried during the summer.

103
Matt Munro | Lonely Planet

Bluebell wood › Hampshire, England The deciduous woods of southern England are filled with the scent of these graceful, dark-blue flowers in May. England has around half the world's population of bluebells.

104
David Clapp | Getty Images

Ceann Hulavig › Isle of Lewis, Scotland More beautiful than Stonehenge and with less restricted access, the Callanish stone circle and its neighbour Ceann Hulavig is the perfect place to celebrate the summer solstice. It was constructed almost 5000 years ago.

106
Helen Hotson | Shutterstock

Mani Rimdu festival › Sagarmatha, Nepal During this 19-day Buddhist festival, held across Himalayan Nepal, monks perform rituals before a three-day public festival begins. It takes place in the 10th month of the Tibetan lunar calendar.

107
Richard I'Anson | Getty Images

Cherry blossom › Yuantouzhu, China It's not just Japan that reveres cherry blossom; when hundreds of the trees burst into flower on Yuantouzhu, a peninsula at Lake Tai, visitors arrive to savour the sight.

108
200 | Getty Images

San Andrés Apóstol cemetery › Mixquic, Mexico The Day of the Dead, a blend of Catholicism and ancient Aztec rituals, is a joyous occasion, celebrating the lives of loved ones. It takes place in early November.

110
Roberto Michel | Shutterstock

Wandering albatrosses › South Georgia Island, Atlantic Ocean Albatrosses tend to mate for life, returning every year of their life – which may be as long as 30 years – to the same colony in the South Atlantic.

111
MZPHOTO.CZ | Shutterstock

Holi festival › India The Hindu spring festival Holi takes place with an explosion of colour across India, but especially in Gujarat. It celebrates the fertility of the land.

112
Yavuz Sariyildiz | Shutterstock

The Sardine Run › Eastern Cape, South Africa When billions of sardines spawn then swim north up the South African coast, predators such as these Common dolphins arrive to pick up an easy meal.

114
Wildestanimal | Shutterstock

Lantern festival › Chiang Mai, Thailand Yi Peng, Chiang Mai's version of the Thai festival Loi Krathong, takes place during a full moon in November and sees lanterns launched into the night sky.

116
Athit Perawongmetha | Getty Images

Snow geese › Canada Snow geese fly south from Canada to pass the winter at Bosque del Apache National Wildlife Refuge on the Rio Grande.

118
FotoRequest | Shutterstock

Polar bear › Svalbard, Norway Pregnant polar bears retreat to a den to give birth during winter, but most polar bears shake off winter by entering a state of 'walking hibernation' when they may not eat for weeks.

119
Ondrej Prosicky | Shutterstock

Vineyards in Greve › Tuscany, Italy Romans, and before them the Etruscans, cultivated vines like these in Chianti. Wine wasn't reserved for special occasions; a daily allowance was recommended.

120
Dan74 | Shutterstock

Danum Valley Conservation Area › Borneo, Malaysia This important parcel of lowland forest in Sabah is home to some of Malaysia's rarest creatures, including orang-utans and pygmy elephants.

122
Nokuro | Shutterstock

The Austfonna ice cap › Svalbard, Norway High up at the tip of Svalbard, on the edge of the Arctic Ocean, lies the largest ice cap in Europe. It has a maximum thickness of more than 500m, which thins a little during the summer.

125
TobyG | Shutterstock

The Dolomites › South Tyrol, Italy Millions of years ago the pale peaks and pinnacles of the Dolomites lay on the seabed; now they are among the world's most distinctive mountainscapes.

126
Matt Munro | Lonely Planet

Monument Valley Tribal Park › Arizona–Utah, USA The sandstone spires of Monument Valley, part of the Colorado Plateau, are the result of millions of years of erosion. Iron oxide gives the rock its reddish tone.

128
Ron and Patty Thomas Photography | Getty Images

Reed Flute Cave › Guangxi, China The limestone stalactites and stalagmites of this cave in Guilin have formed drip by drip over more than 100 million years. They're illuminated for sightseers.

129
Loco Moco Photos | Getty Images

Yellowstone National Park › Wyoming, USA The colours of the Grand Prismatic Spring, the largest hot spring in the USA, derive from different types of bacteria that each thrive in a certain temperature of water.

130
Noppawat | Getty Images

Francois Peron National Park › Western Australia Western Australia's desert meets the Pacific Ocean at Cape Lesueur. This finger of land protects the fragile coastline lying behind it from erosion.

132
Imagine Earth Photography | Shutterstock

Wildflowers in the Atacama Desert › Chile Even in the driest place on Earth, after rain falls the Atacama Desert blooms in vivid colour. The 'desierto florido' happens every five to seven years.

133
abriendomundo | Shutterstock

The Black Forest › Baden-Württemberg, Germany In the southwest of Germany, winter turns the Black Forest's trees white and cross-country skiers and snowshoe-wearing hikers take to the trails.

134
Patrick Poendl | Shutterstock

Drakes Passage › Southern Ocean, Antarctica Named in English after 16th-century sailor Sir Francis Drake, this is the tumultuous stretch of water between Antarctica and South America; it was closed until 41 million years ago.

135
Mike Hill | Getty Images

The aurora borealis › Reine, Norway When charged particles, which flow from the sun at 1.4 million km/h, hit the Earth's magnetic field at the planet's poles, they create curtains of light. Solar storms heighten the effect.

136
Justin Foulkes | Lonely Planet

Zhangye Danxia Landform Geological Park › Gansu, China In northwest China, the colourful stripes of this rock formation come from the red sandstone and minerals, which date back 24 million years.

138
Bule Sky Studio | Shutterstock

Fall leaves › Vermont, USA The trees of the northeast USA have some of the most spectacular displays of autumn foliage when chlorophyll retreats from their leaves as days get colder and darker.

139
SNEHIT PHOTO | Shutterstock

Bora Bora › French Polynesia, Pacific Ocean Once an extinct volcano, Bora Bora is the most famous of the Leeward Islands in the South Pacific. Tropical fish swim in the lagoon formed by its outer rim.

140
theislandexplorers.com | Shutterstock

A supercell storm near Severy › Kansas, USA Supercell thunderstorms have a powerful, rotating updraft. They're prevalent across a strip of the USA's Great Plains known as Tornado Alley.

142
GSW Photography | Shutterstock

Horseshoe Bend › Arizona, USA At Horseshoe Bend, near the town of Page, the Colorado River has carved an 'entrenched meander' through the softer sandstone Eventually it will form a natural bridge.

143
Matt Munro | Lonely Planet

Iceberg in Grandidier Channel › Pleneau Island, Antarctica Although Grandidier Channel is one of the navigable channels of Antartica, mariners have to beware of icebergs, 40,000 of which are calved every year.

144
Chonnie | Shutterstock

Slot canyon › Utah, USA Utah's sandstone plateaus are sliced by thousands of slot canyons, formed when a crack in the rock is forced wider by swirling water, eddying around imperfections.

145
Michelle McCarron | Getty Images

Eyjafjallajökull volcano › Iceland Electrical storms are not uncommon when a volcano erupts explosively – they are generated by charged particles of volcanic ash.

146

Arctic-Images | Corbis

Grand Canyon › Arizona, USA The current Grand Canyon is just six million years old. As canyons age they get broader and deeper; the Grand Canyon is getting deeper each year by the thickness of a sheet of paper.

147

franckreporter | Getty Images

Skógafoss › Iceland This is one of the largest of Iceland's many waterfalls, thanks to the island's abundant rain and rivers. The spray from Skógafoss often creates a rainbow.

148

Peerakit JIrachetthakun | Getty Images

Camel caravan › Sahara, Morocco West Africa's Tuareg traders run camel caravans during the winter through the Sahara Desert, carrying salt and enough food for the camels.

150

MintImage | Shutterstock

Lighthouse Reef › Belize At 125m deep, Belize's Great Blue Hole isn't bottomless but the sinkhole is connected to an extended series of underwater caves formed in the limestone tens of thousands of years ago.

152

Schafer & Hill | Getty Images

Milford Sound › Fiordland National Park, New Zealand This is a place of superlatives, where the rainfall is measured in metres and the waterfalls are among the world's tallest. Milford Sound's walls extend 1200m upwards.

153

Frans Lemmens | Corbis

Hang Son Doong cave › Phong Nha-Ke Bang National Park, Vietnam Hang Son Doong (Mountain River Cave) is known as the world's largest cave, with access only approved by the Vietnamese government in 2013.

154

Mike Rowbottom | Getty Images

Seljalandsfoss › Iceland There are many waterfalls in Iceland but Seljalandsfoss's claim to fame is that visitors can venture behind it. It is just off the southern Ring Road.

156

Pétur Reynisson | Getty Images

A cenote near Valladolid › Yucatán, Mexico There are thousands of these water-filled, subterranean chambers in the Yucatán; the water comes from rain slowly filtering through the ground.

157

better world_10 | Shutterstock

Meteorite crater at Gosse Bluff › Northern Territory, Australia Thought to be the result of an asteroid impact in the heart of Australia, just west of Alice Springs, the edges of this crater have been eroded over time.

158

Stephan Fischer | Shutterstock

Picos de Europa › Asturias, Spain The Picos de Europa in northwest Spain offer some of Europe's finest walking country. At the higher elevations, meadows are backed by a landscape of lakes and limestone peaks.

160

Justin Folkes | Lonely PLanet

Hammerhead sharks › Galápagos Islands, Ecuador Under the oceans' surface, strange phenomena can be seen, including large gatherings of Scalloped Hammerhead sharks. At night the sharks disperse to forage for food.

162

Janos Rautonen | Shutterstock

Edgerton Highway › Alaska, USA The Edgerton Highway is a short 33-mile (53km) extension from the Richardson Highway in Alaska. The Richardson connects Valdez to Fairbanks and opened up a fraction of Alaska's interior.

164

Michael Heffernan | Lonely Planet

Oneonta Creek, Columbia River Gorge › Oregon, USA So narrow that hikers wading the creek can touch both walls, the Oneonta Gorge is home to a variety of ferns and mosses that grow only in this corner of Oregon.

165

zschnepf | Shutterstock

Victoria Harbour › Hong Kong Hong Kong is one of the most densely populated cities in the world and as a result it has one of the highest rates of public transport usage in the world.

166

Adrienne Pitts | Lonely Planet

McDonald Observatory › Texas, USA Far from the light pollution of big cities, West Texas boasts some of North America's clearest and darkest skies, making it the perfect spot for an observatory. Join a star party here to view space.

168

Kris Davidson | Lonely Planet

Sunset › Antarctic Peninsula No place on Earth compares to this vast white wilderness. It's the highest, driest and coldest of the continents and larger than North America. It contains 90% of Earth's ice, which is in the process of melting.

169

Steve McClanahan | 500px

White Sands National Park › New Mexico, USA One of America's three newest national parks, White Sands was once covered by a sea that shed its minerals as it receded. Today, the gypsum dunes can be explored (except when missiles are being tested nearby).

170

Justin Foulkes | Lonely Planet

Sossusvlei sand dunes › Namibia The older the sand dunes in the Namib-Naukluft National Park, the redder they are. The Sossusvlei's dunes can reach 200m in height and they collect dew from the neighbouring ocean.

171

Stephen Barnes | Shutterstock

Hadrian's Wall › Northumberland, England Built in AD122 to keep barbarous northerners out of Roman Britain, Hadrian's Wall was one of the most fortified borders of the time. Today it lies entirely within England.

172

Justin Foulkes | Lonely Planet

Cypresses in the Val d'Orcia › Tuscany, Italy Etruscans planted cypresses (actually from Persia or Syria) around their burial grounds because the trees kept their leaves in winter and were fragrant.

174

Peter Zelei | Getty Images

Pura Ulun Danu Bratan temple › Bali, Indonesia Built in the 17th century to honour the water goddess Dewi Danu, this is one of Bali's most important temples. Traditionally, Bali has relied on irrigation for its agriculture.

176

Guitar photographer | Shutterstock

The Devil's Marbles in Tennant Creek › Northern Territory, Australia There are balancing stones all over the world, from Hampi in India to the southwest of the USA. These ancient stones in Australia are formed from granite that has weathered.

177

totajla | Shutterstock

Central Park in Manhattan › New York, USA During the 19th century, Central Park was carefully sculpted; swamps were drained and topsoil imported from New Jersey. It wasn't until the 1960s that cars were banned from the park at weekends.

178

Lottie Davies | Lonely Planet

Red-crowned cranes › Hokkaido, Japan The rare red-crowned cranes of Hokkaido, of which there are about 1000, dance in pairs throughout the year, reinforcing the pair's bonds. The local Ainu people call them 'the gods of the marshes'.

180

Pichit Tongma | Shutterstock

Red-crowned cranes › Hokkaido, Japan The cranes' dances include bowing, jumping, synchronised strutting and the tossing of grass and sticks. To the Japanese the cranes symbolise luck and longevity.

181

Piterpan | Getty Images

Triglav National Park › Slovenia Slovenia's only national park fans out around the country's highest peak, Mt Triglav. This land of forests and crystalline lakes in the Eastern Julian Alps is bordered by Italy and Austria.

182

Andrew Mayovskyy | Shutterstock

Reindeer herding › Sweden The Sámi people work on the land that spans northern Norway, Sweden and Finland. They still live in 'siiddat' (reindeer herding groups) and use the animals for transport, milk and meat.

184

Gary Latham | Lonely Planet

Mustering horses › Xilin Gol, Mongolia Mongolia's xilingol horse is not wild but the breed is at home on the vast grasslands and is used for riding and work.

185

HuanPhoto | Shutterstock

Tsingy de Bemaraha Strict Nature Reserve › Madagascar Animals – including lemurs such as the sifaka – and plants have evolved to live among the jagged pinnacles (tsingys) of this park in the west of Madagascar. The reserve was founded in 1998.

186

Justin Foulkes | Lonely Planet

Moraine Lake in Banff National Park › Alberta, Canada Moraine Lake, fed by glacial waters, is overlooked by the Valley of the Ten Peaks. The peaks range from 3000m to 3400m. Mt Allen is named after the cartographer who mapped the region.

187

Matt Champlin | Getty Images

Highland cattle › Scotland Scotland's distinctive Highland cattle were bred to survive cold weather and to forage for food in a harsh landscape. Despite appearances, they're very mild-mannered.

188

Daniel Alford | Lonely Planet

The Cuillins › Isle of Skye, Scotland The wild Black Cuillins of this island in the Inner Hebrides draw climbers and hikers from far and wide. The island is home to golden eagles and red deer.

189

Daniel_Kay | Shutterstock

The Kimberley coast › Western Australia The coastline of this vast and hitherto unspoiled wilderness extends 1300km and is home to the world's largest population of humpback whales.

190

Matt Deakin | Shutterstock

A quiver tree › Kalahari, Namibia–South Africa The quiver tree survives in its arid world by storing water in thick green leaves and fat white branches. It can self-amputate branches when conditions are exceptionally dry.

192

Jaco Wolmarans | Getty Images

Kamakura Bamboo Garden › Tokyo, Japan Formal Japanese gardens are often designed for Zen-like meditation, with each plant selected for its aesthetic appeal; bamboo is sometimes used to hide features and create a sense of mystery.

193

Marco Maccarini | Getty Images

Lamayuru monastery › Ladakh, India One of the oldest gompas (monasteries) in Western Ladakh, Lamayuru lies on the Srinagar–Leh road in Kargil district.

194

Jaturong Kengwinit | Getty Images

Rice paddies › Bali, Indonesia Terraced fields allow rice to be grown on steep slopes and make irrigation easier. Ubud in central Bali is famed for this type of farming.

196

John Laurie | Lonely Planet

Tianmen mountain road › Hunan, China It took eight years to build this 10km stretch of road in the Tianmen Mountain National Park, which is in the southeast of the country.

197

THONGCHAI.S | Shutterstock

Wildebeest migration › Masai Mara, Kenya At 800km, it's not the world's longest migration but it is the largest in volume. Two million wildebeest move from the Serengeti to the Masai Mara, braving the crocodiles of the Mara River along the way.

199

Gudkov Andrey | Shutterstock

Victoria Falls › Zambia–Zimbabwe At more than 1700m in width, Victoria Falls have the highest recorded volume of water tumbling over the edge: 12,800 cu metres per second. Peak flow is in April.

200

Wolfgang_Steiner | Getty Images

Bryce Canyon National Park › Utah, USA The Wall Street Trail of Bryce Canyon has its own skyscrapers: ponderosa pines. Chasms in the rock are formed when water freezes and expands, creating alleys up to 60m deep.

202

Sarun Laowong | Getty Images

The Gasherbrum massif › Karakorum, Pakistan The remote Gasherbrum massif on the border of Pakistan and China has one of the highest concentrations of unclimbed peaks of any mountain range in the world.

203

NG-Spacetim | Shutterstock

The Great Wall › China The Great Wall is a conjunction of walls, initially constructed for defence and then to control immigration and levy taxes. The wall, restored by the Ming dynasty, extends 8850km.

204

Mark Read | Lonely Planet

Mount Bromo › East Java, Indonesia Rising from the ruins of the ancient Tengger caldera, Gunung Bromo is one of three volcanoes to have emerged from a vast crater. Flanked by the peaks of Kursi and Batok, Bromo stands in a sea of volcanic sand.

205

Sugrit Jiranarak | Shutterstock

Angkor Wat. Siem Reap, Cambodia › Angkor Wat is where the civic and the spiritual meet. The world's largest religious monument, 'the temple that is a city' was built by Khmer kings in the 12th century.

206

Mark Read | Lonely Planet

Baobab tree › Sri Lanka › Baobab trees can grow all over Asia, Africa and Australia. This one, in Sri Lanka, is reputed to have sprung from a cutting from the tree under which Buddha once sat in India.

207

Jonathan Stokes | Lonely Planet

Los Glaciares National Park › Patagonia, Argentina Mt Fitzroy to the right and spiky Cerro Torre to the left are the two landmark mountains in Patagonia's ice fields, the world's largest outside Antarctica.

208

Nido Huebl | Shutterstock

Whale shark › Western Australia The world's largest fish species has a particular fondness for Ningaloo Reef, off Western Australia, where it filters the water for plankton and tiny fish.

210

soft_ligh | Shutterstock

Ningaloo Marine Park › Western Australia At places, it's possible to swim from the shore of Western Australia to the world's largest fringing reef, the Ningaloo, which is home to more than 500 species of fish, turtles and rays.

211

Violeta Brosig | Shutterstock

The Palace of Westminster › London, England After a fire that razed the Houses of Parliament in 1834, Sir Charles Barry redesigned the Palace of Westminster. Big Ben is the name of the bell, which tolls at the top of Elizabeth Tower.

212

Enzo Figueres | Getty Images

Monument Valley › Arizona–Utah, USA Monument Valley's fame far exceeds its size thanks to roles in numerous movies. The 300m-high sandstone buttes have appeared in films as diverse as Stagecoach and 2001: A Space Odyssey.

214

ronnybas | Shutterstock

Elephants › Okavango Delta, Botswana Most rivers empty into an ocean but not landlocked Botswana's Okavango river. It floods a plain of 5000–15,000 sq km, creating a watery haven for wildlife large and small.

216

Zaruba Ondrej | Shutterstock

Millau Viaduct › Midi-Pyrénées, France The world's tallest bridge spans the valley of the River Tarn in southern France for 2460m. It cost 400 million Euros to construct and is estimated to last 120 years.

218

FraVal Imaging | Shutterstock

Yosemite National Park › California, USA North America's highest measured waterfall, Yosemite Falls, cascades into the Merced . River. Yosemite National Park covers 3000 sq km, though it is Yosemite Valley that draws the largest number of visitors.

219

Mark Read | Lonely Planet

Strokkur geyser › Iceland The Strokkur geyser erupts every four to eight minutes, blasting water up to 40m into the air. The word 'geyser' itself comes from Icelandic, 'geysa', which means 'to gush'.

220

Francesco R. Iacomino | Shutterstock

Limestone pinnacles at Wulingyuan › Hunan, China The karst shards of Wulingyuan are what remain of quartzite sandstone mountains after millions of years of water erosion. They're part of Zhangjiajie National Forest Park.

221

Feng Wei Photography | Getty Images

Angel Falls › Venezuela Jimmie Angel was an American aviator who flew over the world's tallest waterfall in 1933. He landed there in 1937 with his wife but his plane became stuck; they trekked for 11 days to reach safety.

223

James Marshall | Corbis

The Royal Tomb at Petra › Jordan Petra's significance is due to its historic location between Arabia, Egypt and Syria-Phoenicia. The crossroads of trading routes was inhabited since prehistoric times.

224

Wallacefsk | Getty Images

Teahupoo › Tahiti The most famous break in French Polynesia is also one of the biggest, heaviest waves in the world to surf, which crashes on a shallow, razor-sharp reef. Experts only!

226

vladimir3d | Shutterstock

The Wave, Coyote Buttes › Arizona–Utah, USA Two effects are on display at this part of the Paria Canyon–Vermilion Cliffs Wilderness: the laying down of sediment under long-gone seas, and the wearing down of rock by the elements.

227
Johnny Adolphson | Shutterstock

Uluru › Northern Territory, Australia Uluru is 335km southwest of Alice Springs in Australia's hot, red heart. It's formed from rock thought to have been deposited more than 500 million years ago, and it once lay on a sea's bed.

228
Michael Dunning | Getty Images

Taj Mahal. Uttar Pradesh › India The Taj Mahal is a monument to love, composed in white marble – over 20 years from 1632 – to honour Mumtaz Mahal, wife of Mughal Emperor Shah Jahan.

229
Frans Lemmens | Getty Images

Zabriskie Point › Death Valley. California, USA Death Valley was the bed of a lake millions of years ago. Its sediment created the stripes seen at Zabriskie; the dark layer is from five-million-year-old volcanic eruptions.

230
Stephanie Sawyer | Getty Images

Terracotta Army, Xi'an › Shaanxi, China The Terracotta Army was buried with Emperor Qin Shi Huang in 210BC. The 8000 soldiers were to protect the emperor in the afterlife. He was also buried with musicians and acrobats.

232
Bule Sky Studio | Shutterstock

Terracotta Army, Xi'an › Shaanxi, China The figures in the Terracotta army were constructed in pieces, then assembled. They have different facial expressions and vary in height and uniform; their weapons were real.

233
Intarapong | Shutterstock

Shiprock › New Mexico, USA Rather than being a sandstone buttress, Shiprock is actually the eroded vestige of a volcanic plug, standing 2188m above the desert. The Navajo people wove many myths about its form.

234
Justin Foulkes | Lonely Planet

The Temple of Saturn › Rome, Italy Standing in ancient Rome's Forum, this temple was dedicated to Saturn, the scythe-wielding Roman god of agriculture. Saturnalia was Rome's most anticipated festival, and the god also lent his name to Saturday.

236
Justin Foulkes | Lonely Planet

Bristlecone pine › California, USA Bristlecone pine trees, like this one in the Inyo National Forest of California's Eastern Sierra, are believed to live longer than any other organism – up to 5000 years.

237
Bill45 | Shutterstock

Kluane Glacier › Yukon, Canada Glaciers grind their way surprisingly quickly through the St Elias mountains, where many of North America's 16 highest peaks lie. A warming planet means the glaciers are less stable.

238
Justin Foulkes | Lonely Planet

Bondi › New South Wales, Australia Famous, sun-bronzed Bondi beach is the poster child for fun-filled good times. But even this place is connected to the rest of the planet as swells hit the sand having travelled across oceans.

239
Jonathon Stokes | Lonely Planet

Bagan temples › Mandalay, Myanmar Bagan was the capital of a pre-Myanmar kingdom; it was both a spiritual place, with 2000 temples remaining from 10,000, and a secular place of scholarship.

240
Ikunl | Shutterstock

West Elk Wilderness › Colorado, USA The cycles of the moon, including eclipses, have been predicted for thousand of years to come. The pattern is dependent on the movements of the sun, the moon and Earth.

241
Bridget Calip | Shutterstock

Rock Islands Southern Lagoon. Palau, Micronesia The 445 uninhabited limestone or coral islets in the western Pacific Ocean were made a Unesco World Heritage Site in 2012 due to their biodiversity and human history.

242
Bob Krist | Corbis

San Rafael waterfall. Ecuador › The Coca River cascades 150m to create Ecuador's largest waterfall. It lies in the important ecosystem of the Sumaco Biosphere Reserve, where the Andean and Amazonian regions meet.

Front cover
Shobeir Ansar | Getty Images

Beautiful World
April 2021
Published by Lonely Planet Global Limited
CRN 554153
www.lonelyplanet.com
10 9 8 7 6 5 4 3 2 1

Printed in China
ISBN 978 1838 69467 8
© Lonely Planet 2021
© photographers as indicated 2021

Managing Director, Publishing Piers Pickard
Associate Publisher & Commissioning Editor Robin Barton
Designer Jo Dovey
Image re-touching Ryan Evans and Jo Dovey **Writer** Kate Morgan
Print Production Nigel Longuet

Lonely Planet Offices

Ireland
Digital Depot, Roe Lane (off Thomas St),
Digital Hub, Dublin 8,
D08 TCV4

USA
230 Franklin Road, Building 2B,
Franklin, TN 37064
T: 615-988-9713

STAY IN TOUCH lonelyplanet.com/contact

Paper in this book is certified against the Forest Stewardship Council™ standards. FSC™ promotes environmentally responsible, socially beneficial and economically viable management of the world's forests.

LONELY PLANET'S
BEAUTIFUL WORLD